Bibliographic information published by the German National Library:

The German National Library lists this publication in the National Bibliography; detailed bibliographic data are available on the Internet at http://dnb.dnb.de .

Imprint:

Print and binding: Books on Demand GmbH, Norderstedt Germany
ISBN: 9783668926387

This book at GRIN:

https://www.grin.com/document/463480

Dominik Ohlmann

Novel Hydroxamic Acid-Derived Ligands for Vanadium Catalysed Asymmetric Epoxidations

GRIN Verlag

Novel Hydroxamic Acid-Derived Ligands for Vanadium Catalysed Asymmetric Epoxidations

Dominik Ohlmann

ERASMUS Exchange Project Report

April – June 2007

Department of Chemistry
University of Glasgow

ACKNOWLEDGEMENTS

First of all I would like to thank my supervisor Dr Andrei Malkov for his friendly tutelage, the interesting project, and the opportunity to work in his group.

Special thanks to Louise Czemerys for her friendly advice in the laboratory, her patient helpfulness, and finally for reading this report and correcting my English.

Many thanks go to all the other group members of laboratory A4-31 and A4-32 in Joseph Black Building for their steady assistance and support in everyday laboratory work, and for their open and friendly demeanour: Aneta, Claire, Frédéric, Grant, Kveta, Jan, Marek, Mikhail and Sigitas.

ABBREVIATIONS

anh.	anhydrous
aq.	aqueous
bp.	boiling point
equiv.	equivalent(s)
mg	milligram(s)
h	hour(s)
min	minute(s)
mL	millilitre(s)
mmol	millimole(s)
r.t.	room temperature
M	mol l^{-1}
temp.	temperature
acac	acetylacetonato
Boc	*tert*-butyloxycarbonyl
Bpoc	2-biphenyl-4-ylisopropyloxycarbonyl
t-BuOOH	*tert.*-butyl hydroperoxide
m-CPBA	*meta*-chloroperbenzoic acid
DCM	dichloromethane
DIEA	diisopropylethyl amine (Hünig's base)
DMF	*N,N*-dimethyl formamide
Dpp	*P,P*-diphenylphosphino
Et_2O	diethyl ether
EtOH	ethanol
Et_3N	triethyl amine
HCl	hydrochloric acid
H_2SO_4	sulfuric acid
KOH	potassium hydroxide
Ment	(*L*)-menthyl
$MgSO_4$	magnesium sulfate
Na_2CO_3	sodium carbonate
NaH	sodium hydride

$NaHCO_3$	sodium hydrogen carbonate
$NaNO_2$	sodium nitrite
Na_2SO_3	sodium sulfite
NMM	*N*-methyl morpholine
Nosyl	*ortho*-nitrophenylsulfonyl
PE	petroleum, bp. 40-60 °C
Ph	phenyl
TFA	trifluoroacetic acid
THF	tetrahydrofuran
TLC	thin-layer chromatography
$VOSO_4 \cdot H_2O$	vanadyl sulfate hydrate

GC	gas chromatography

HPLC	high performance liquid chromatography

HRMS	high-resolution mass spectrometry
m/z	mass per charge

NMR	nuclear magnetic resonance
s	singlet
d	dublet
t	triplet
bs	broad singlet
m	multiplet

TABLE OF CONTENTS

1 Introduction

1.1 Sharpless Epoxidation

Asymmetric epoxidation of olefinic double bonds has the potential to create two chiral centres in one step. Many research efforts have therefore been dedicated to the development of chiral metal catalysts that can perform asymmetric epoxidation.[1] Most well-known is probably the *Sharpless* method, based on titanium tartrate complexes, for epoxidation of allylic alcohols. It was discovered in 1977 that a combination of $VO(acac)_2$ and chiral hydroxamic acid (L^* in Scheme 1) afforded optically active epoxides from the corresponding allylic alcohols and *t*-butyl hydroperoxide as oxidant with up to 50% ee. With a proline-derived hydroxamic acid as chiral ligand, higher enantioselectivities (up to 80% ee) could be achieved by the same authors a few years later.[2]

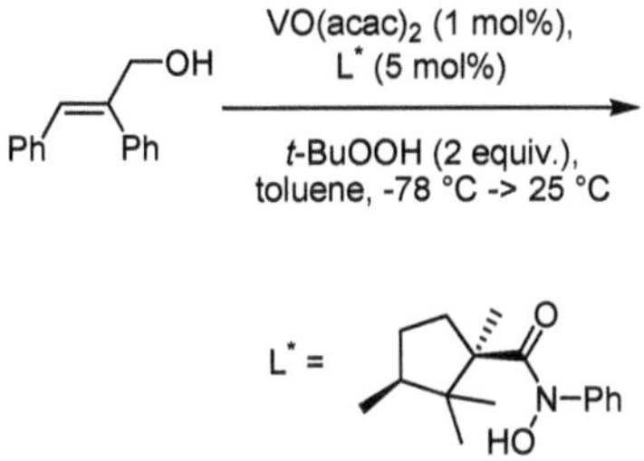

Scheme 1. Asymmetric epoxidation according to *Sharpless.*[3]

Although this method is well understood and capable of delivering reliable and reproducible results, it still shows some disadvantages such as high catalyst loading, a tedious work-up and furthermore requires stringently anhydrous conditions.

1.2 Yamamoto-Malkov Epoxidation

In 2000, *Yamamoto*[4] reported the successful use of amino acid-derived phthalimide ligands (**A** in Figure 1) in vanadium(V) systems which came close to Ti(IV) catalysts concerning reactivity and enantioselectivity. At the same time, *Malkov*[5,6] showed that phenylglycine derived sulfonamide ligands (**B** in Figure 1) exhibited high reactivity and promising enantioselectivity as well. But still, in both cases toluene was used as solvent and reactions had to be carried out under anhydrous conditions.

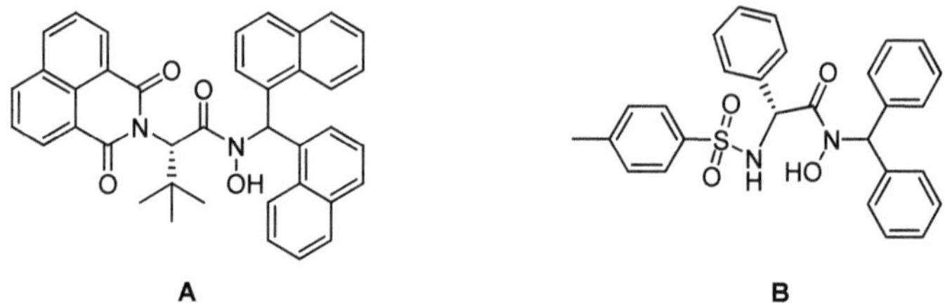

Figure 1. Selected hydroxamic acids used as ligands so far.

Then it was discovered by the *Malkov* that the epoxidation reaction (Scheme 2) can successfully be carried out in water[6], where the costly vanadium precursor $VO(i\text{-}OPr)_3$ and anhydrous *t*-BuOOH can be replaced with much cheaper alternatives, such as $VOSO_4$ and aqueous *t*-BuOOH.

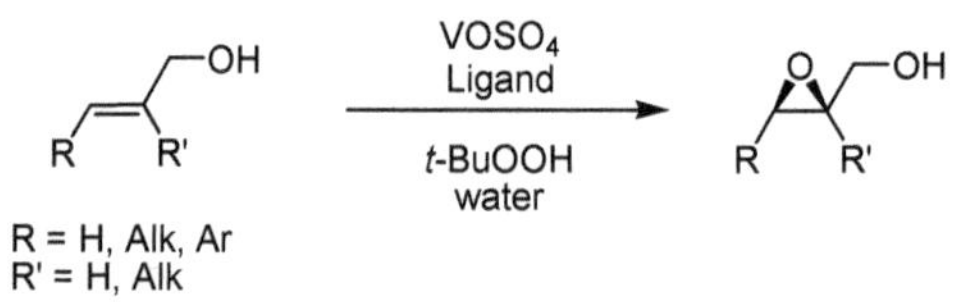

Scheme 2. Asymmetric epoxidation in water according to *Malkov*.

The reaction rate in water turned out to be generally slower than in organic solvents, and the *Yamamoto* ligand resulted in low conversions. However, the sulfonamides provided a reasonable level of reactivity, and enantiomeric excess values on a range of epoxidized allylic alcohols achieved up to 72 %.[6] Another advantage of the application to aqueous systems consists of the very low catalyst loading needed (1 mol% or less).

So far, the coordination of ligands had lead to a significant deactivation of the catalyst in case of vanadium, resulting in a ligand-decelerating process. Applying the methodology mentioned above, it has been found that in water, vanadium catalysis turns into a ligand-accelerated process. Regarding the ligands, this effect can be attributed to the hydrogen bonding between the sulphonamide moiety and the incoming alcohol which brings and holds the reactants together. As well, transition states and active complexes have been studied[6].

1.3 Camphor Derivatives

Chiral non-racemic monoterpenes derived from naturally occurring compounds have been widely used as chiral building blocks for the preparation of auxiliaries for asymmetric synthesis and ligands for asymmetric catalysis. Amongst monoterpenes a prominent position is occupied by optically active camphor, the framework of which has been incorporated into a variety of chiral ligands having homo- or heterodonor atoms.[7] Interest in the chemistry of camphor and its derivatives has been continuous throughout the history of natural product chemistry. This interest is largely associated with the fact that (+)- or (–)-camphor is readily available and undergoes a wide variety of transformations, sometimes involving rearrangement processes. The range of functionalizations of the chiral moiety has been reviewed recently.[8]

Regarding applications in catalytic asymmetric synthesis, a class of camphor-based pyridine ligands have recently gained interest caused by improved methodology related to allylic substitution, allylic oxidation, cyclopropanation and hydrogenation[7] (**A**, Figure 2).

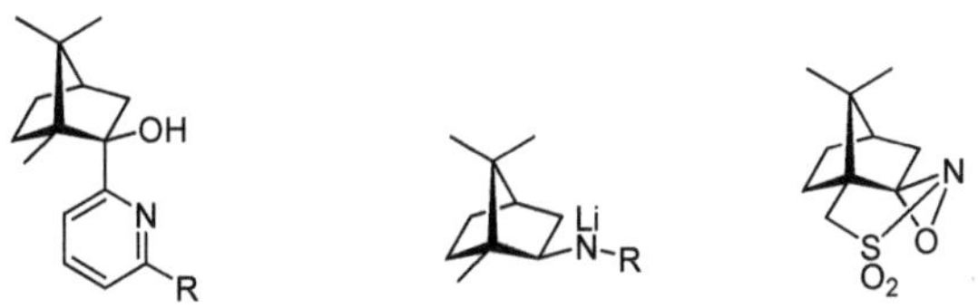

A
R = H, 2-pyridyl, phenyl

B
R = COPh, Ph

C

Figure 2. Selected camphor-based ligands for asymmetric synthesis.

Further applications of camphor-derived catalysts can also be found concerning carbonyl reductions, enantioselective deprotonations and α-hydroxylations[9] (**B** and **C**, Figure 2), proving that camphor derivatives provide a valuable asset for reliable and versatile preparations of enantiopure compounds.

1.4 Phosphinamides

The chemistry of phosphinamides first gained interest in1976, as they were used as amino protecting groups in peptide synthesis by *Kenner*, *Moore* and *Ramage*.[10] They designed a series of derivatives bearing similar reactivities as the well-known urethane protecting groups Boc (*tert.*-butyloxycarbonyl) and Bpoc (2-biphenyl-4-ylisopropyloxycarbonyl), which can be cleaved under mildly acidic conditions (Figure 3). For example, Dpp (Diphenylphosphino) is slightly more acid labile than Boc protecting group.[11]

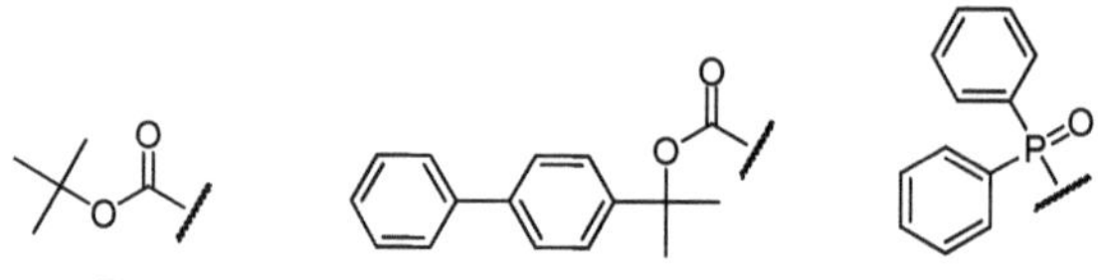

Figure 3. Different acid labile protecting groups.

To explain the particular acid lability of phosphinamides, the authors investigated the stereoelectronic properties of phosphinamides in comparison with amides. It was found that the nitrogen geometry in $Ph_2P(O)NMe_2$ and $Ph_2P(O)N(Me)CH_2CH_2Ph$ shows a flattened tetrahedron in which the lone pair on N is almost in the N-P-O plane, whereas in the amide bond, the lone pair of electrons on the trigonal N is orthogonal to the N-C-O plane.

The Dpp moiety turned out to be suitable as an amino protecting group even in oligopeptide synthesis and was stable during customary manipulations, which were found by the same authors a few years later.[12]

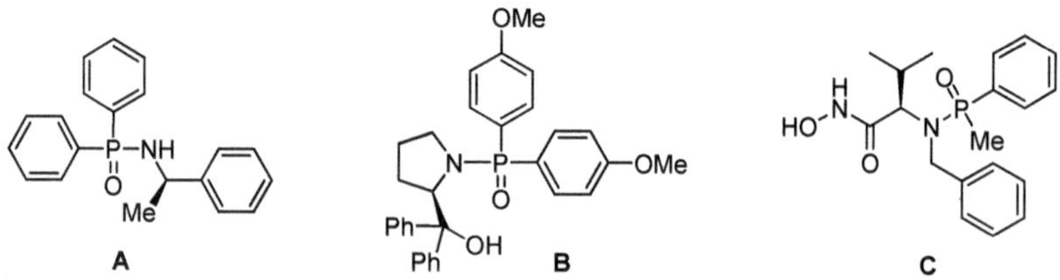

Figure 4. Examples for phosphineamide-based ligands.

Nowadays, aside from its use as protecting group, phosphinamide moiety is mainly used in chiral ligands for asymmetric catalytic reactions, such as the asymmetric borane reduction of prochiral ketones[13,14,15] (ligands **A, B** in Figure 4). Phosphinamide-based hydroxamic acids, which are considered in this work later on, have also found applications as inhibitors of matrix metalloproteinases[16] (**C** in Figure 4).

1.5 Sulfinamides

A relatively rare class of ligands is that in which the stereogenicity resides not at the carbon atoms, but at a heteroatom such as sulfur. Over the last three decades, more than 40 different classes of chiral sulfur compounds have been described in the literature, accompanied by the development of a broad variety of useful procedures for the synthesis of enantiomerically pure sulfur compounds.[17] Transition-metal complexes with chiral sulfur ligands are powerful catalysts in a considerable number of reactions, although, there have been several difficulties concerning chiral sulfinyl compounds, such as the formation of diastereomeric mixtures and the control of asymmetric induction in catalytic reactions.[18] An appropriate summarizing review has recently been given by *Fernandez* and *Khiar*.[19]

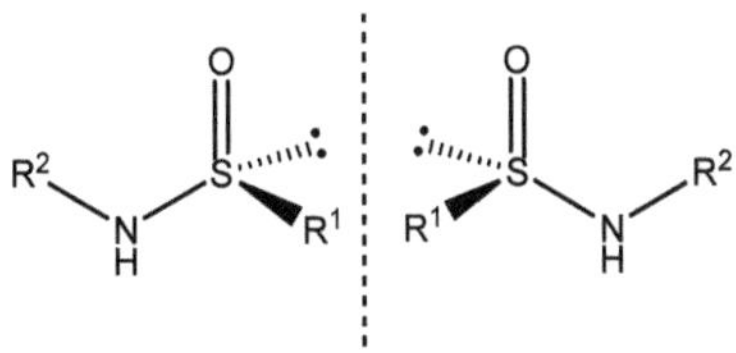

Figure 5. Illustrating the stereogenicity of sulfinamide moiety.

The structural key motif of sulfinamides consists of the sulfur atom in oxidation state (+II), which can also be found in sulfinimines, a class of ligands for asymmetric synthesis gaining interest since the 1990s.[20] Considering the lone pair of sulfur, the structure can formally be regarded as tetrahedral (Figure 5), with activation parameters for pyramidal inversion of $\Delta H^{\#} = 35\ldots42$ kcal mol^{-1} and $\Delta S^{\#} = 8$ kcal K^{-1} mol^{-1} for related chiral sulfoxides. Thermal stereomutation can therefore be achieved at a fairly high rate at roughly 200 °C.[21]

From the view of metal complexation, chiral sulfinamide-containing scaffolds bear attractive features because of their ability to bind through nitrogen, sulfur, or oxygen to form a number of uniquely bonded motifs.[22] Sulfinamides can be easily assembled from simple building blocks[23] and are both structurally[24] and optically stable, as mentioned above. Recent developments in improved synthetic approaches deliver a broad variety of possible stereoselective synthesis and reactions[25], as well as successfully applied auxiliaries in catalytic asymmetric processes.[26]

2 Project Objective

The aim of this project was to carry out of the synthesis of three new classes of ligands ligand (Figure 6), each of which bearing different ways of influencing the product formation of vanadium(V)-catalysed asymmetric epoxidation in water. Since the amino acid backbone and the benzhydryl-amino moieties proved their suitability, these features will be preserved in the new ligands while varying the substituents on amide nitrogen.

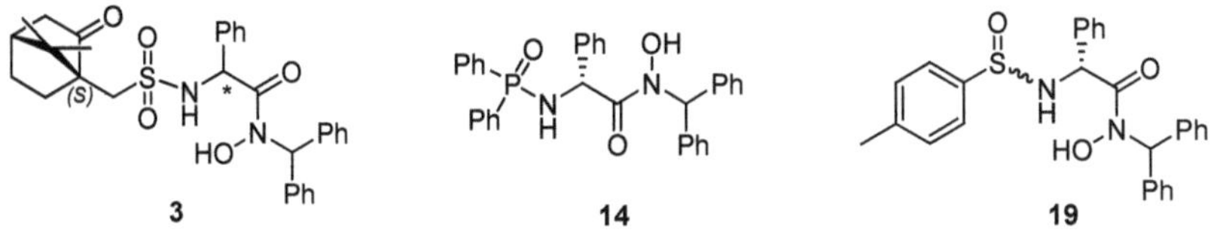

3 14 19

Figure 6. Target ligands of current project.

Design of ligands **3** incorporates the main features of the most successful ligand (**B** in Figure 1) and introduces an additional element of chirality (camphor) to the sulfonamide function. Both diastereomers of **3** are supposed to be prepared and the ligands will be tested in asymmetric epoxidation of model substrates.

Increasing the sterical extent of the amide moiety and, at the same time, exchanging sulfur for phosphorus, led to the development of ligand **14**. Presuming a successful preparation, testing **14** in asymmetric epoxidation will represent the first application of phosphinamide derivatives to this reaction methodology.

In ligand **3**, we have the sulphonamide moiety separating chiral amino acid from the chiral camphor scaffold. Therefore, another aspect one might consider is to decrease the distance of the second centre of stereogenicity to the amino acid backbone. The resulting ligand **19** bears a sulfinamide moiety, which introduces chirality caused by asymmetric substitution of sulfur. Synthesis of **19**, separation of diastereomers and their application to the reaction mentioned above (Scheme 2) is considered as another target of this project.

3 Results and Discussion

3.1 Camphor-derived hydroxamic acids

3.1.1 Synthesis of ligands 3

The use of chiral camphor moiety is the reasonable continuation of the strategy that introduces ligand chirality by modifying the amino acid backbone. Enhancing the, so far non-chiral, sulfonamide moiety by addition of a stereogenic centre should enable the ligand (Figure 7) to influence the stereochemistry of asymmetric epoxidation in a way to reach higher enantiomeric excess. Also, the bulky quality of the bicyclic scaffold is supposed to influence the selectivity of product formation. Furthermore, the sulfonamide moiety is stable against strong acids and bases, and therefore easy to handle and useful substrates can be build even under harsh reaction conditions.

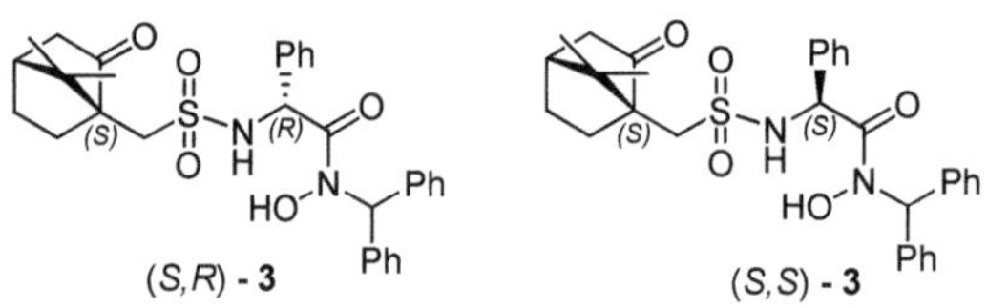

Figure 7. Camphor-derived hydroxamic acid ligands.

Synthesis of each diastereomer (*S*, *R*)-**3** and (*S*, *S*)-**3** was accomplished in four convergent steps (Scheme 3). Starting from enantiopure (*S*)- or (*R*)-phenylglycine, which were coupled under aqueous conditions with commercially available (*1S*)-Camphor-10-sulfonic acid chloride to sulfonamides **1**, the synthesis was continued by using phosphorus pentachloride in anhydrous ether to obtain acid chlorides **2**.

To prevent the possible *O*-acylation in the next step, hydroxylamine **5** was protected *in situ* with chlorotrimethylsilane in presence of 2,6-lutidine, which acted as a non-nucleophilic base.

The subsequent coupling with **2**, carried out in anhydrous THF, and *O*-deprotection with water furnished camphor-derived hydroxamic acids **3** in 44 – 73% yield.

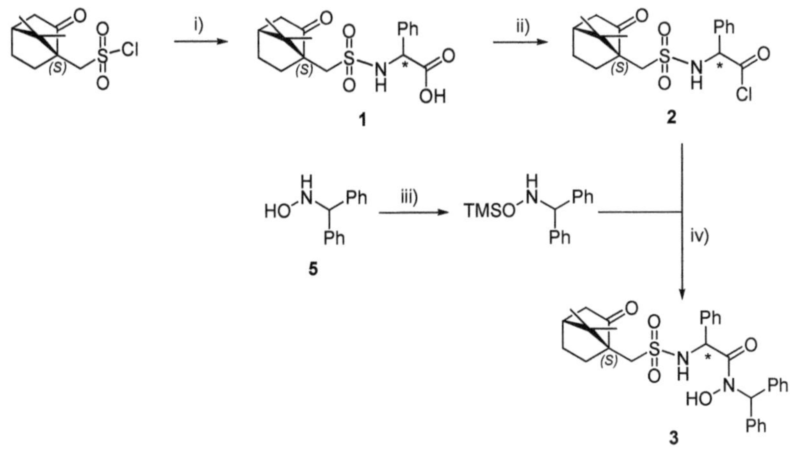

Scheme 3. Synthesis of ligands **3**. Conditions: i) (*S*)- or (*R*)-phenylglycine, aq. NaOH, ether, r.t.; ii) PCl_5, ether, r.t.; iii) chlorotrimethylsilane, 2,6-lutidine, THF, 0 °C; iv) 2,6-lutidine, THF, r.t.

3.1.2 Application of ligand 3 to vanadium-catalyzed epoxidation

Hydroxamic acids **3** were tested in two standard epoxidation runs in water, using allylic alcohols **6a-b** as substrates (Figure 8). The catalyst consisted of vanadyl sulfate hydrate as the metal source and chiral ligand (*S, R*)-**3** or (*S, R*)-**3**. More reaction conditions are given in Scheme 4.

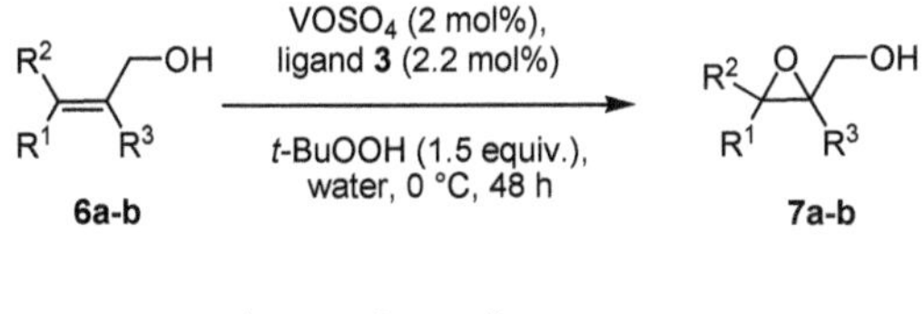

6a, **7a** R^1 = Ph, R^2 = H, R^3 = CH_3
6b, **7b** R^1 = $CH_2CH_2CHC(CH_3)_2$, R^2 = CH_3, R^3 = H

Scheme 4. Asymmetric epoxidation using ligands **3**.

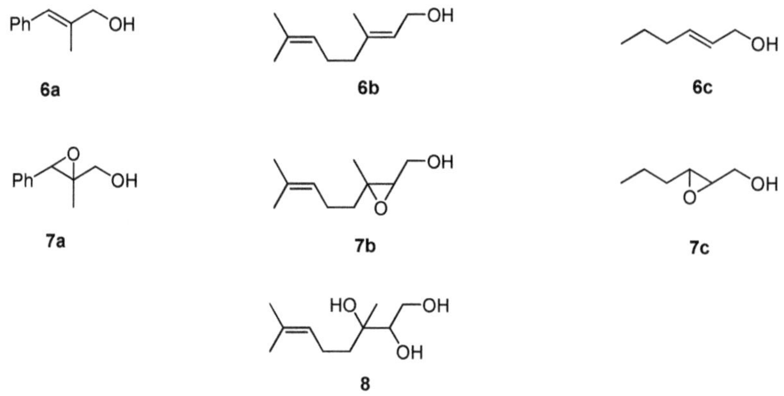

Figure 8. Allylic alcohols used as substrates and appropriate epoxidation products.

Applying the reaction conditions optimized before[6], we obtained moderate enantiomeric excess values (up to 46%) with camphor-derived hydroxamic acids **3** as ligands. The best results could be achieved when using ligand (*S*, *S*)-**3** on substrate **6a** (entry 3 in Table 1). The expected epoxidation product **7b** of geraniol **6b** appears to being opened instantly in presence of ligand (*S*, *R*)-**3**, therefore only the triol **8** could be detected after working up reaction mixture (entry 2 in Table 1).

Since epoxide ring opening occured as well in case of ligand (*S*, *S*)-**3** and substrate **6a**, even though in much lower quantity as with the other ligand diastereomer, the desired epoxide was detected as minor product (entry 3 in Table 1).

entry	ligand	substrate	product	ee (%)
1	(*S*, *R*)-**3**	**6a**	(*R*, *R*)-**7a**	38 [a]
2	(*S*, *R*)-**3**	**6b**	**8**	-
3	(*S*, *S*)-**3**	**6a**	(*S*, *S*)-**7a**	46 [a]
4	(*S*, *S*)-**3**	**6b**	(*R*, *R*)-**7b**	33 [b]

Table 1. Summarized results of epoxidations using ligands **3**. *a* detected by chiral GC (Supelco β-Dex 120 column; oven temp. 110 °C for 2 min, then 1 °C/min to 200 °C, $t_{S,S}$ = 34.66, $t_{R,R}$ = 35.27); *b* Supelco α-Dex 120 column; oven temp. 110 °C for 2 min, then 1 °C/min to 200 °C, $t_{(S,S)}$ = 21.67, $t_{(R,R)}$ = 22.09.

The camphor moiety with its intrinsic chiral scaffold was apparently not able to introduce additional chirality into ligand backbone. It was probably too far away from the coordinating site and therefore not able to exert a sufficient effect on the product formed. The major contribution to enantioselectivity appears to still originate from the chiral amino acid backbone. Also, the camphor moiety's bulky qualities seem to have no significant influence on the selectivity of epoxidation reaction.

3.2 Hydroxamic acid based on diphenylphosphinamide moiety

3.2.1 Synthesis of ligand 14

With the results of camphor derivatives in mind, we considered amplifying the sterical extent of the moiety connected to amide N-H to improve enantioselectivity This could possibly be achieved, for instance, by substituting the Camphor-SO_2 – moiety for Ph_2PO, however, at the expense of chirality on this moiety (Figure 9).

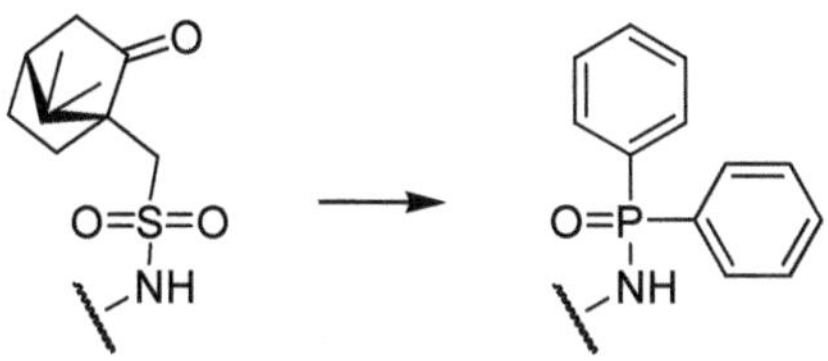

Figure 9. Increasing the sterical extent of amide moiety.

Synthesis of phosphinamide-based hydroxamic acid **14** was accomplished in five convergent steps (Scheme 5). As reported before[12], direct introduction of diphenylphosphino (Dpp) group into free (*R*)-phenylglycine under aqueous or anhydrous conditions (route **a**) is not possible due to extremely quick hydrolysis of DppCl into diphenylphosphinic acid. To prevent DppCl from hydrolyzing, we chose (*R*)-phenylglycine methyl ester hydrochloride as starting material, and under anhydrous coupling conditions *N*-Dpp-amino acid methyl ester **11** was successfully obtained (route **b** in Scheme 5). Subsequent mild alkaline hydrolysis afforded the free acid **12.**

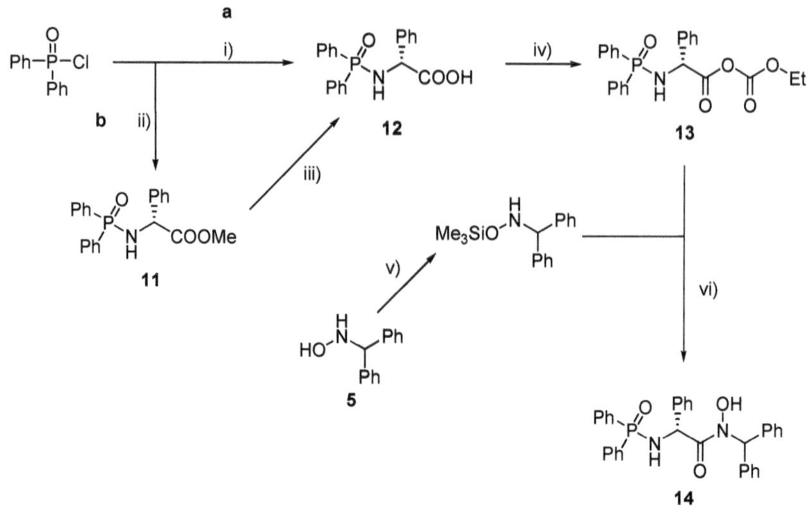

Scheme 5. Synthesis of phosphinamide-based hydroxamic acid **14**. Conditions: i) (*R*)-phenylglycine, anh. THF, r.t.; ii) (*R*)-phenylglycine methyl ester hydrochloride, anh. DCM, 0 °C; iii) DCM, aq. NaOH, r.t.; iv) ethyl chloroformate, NMM, anh. THF, -10 °C; v) chlorotrimethylsilane, 2,6-lutidine, THF, 0 °C; vi) 2,6-lutidine, THF, r.t.

Due to the P-N bond's acid-sensitivity[12] we could not use a chlorination reagent such as $SOCl_2$ or PCl_5 to convert compound **12** into the acid chloride, as we did to generate **2** from corresponding acid **1**. We therefore continued the synthesis via the activated mixed anhydride **13**, which was synthesized from free acid **12** using ethyl chloroformate in the presence of *N*-methyl morpholine at low temperature. Compound **13** was not isolated but immediately used in the following step. There, utilization of previously *O*-protected hydroxylamine trimethylsilyl-**5** in presence of 2,6-lutidine afforded the desired hydroxamic acid **14**.

3.2.2 Application of ligand 14 to vanadium-catalyzed epoxidation

We tested the ligand properties of the phosphinamide-based hydroxamic acid **14** under similar conditions as we did on ligands **3**. This time we used allylic alcohols **6a** and **6c** as substrates (Figure 8) and the reaction conditions given in Scheme 6.

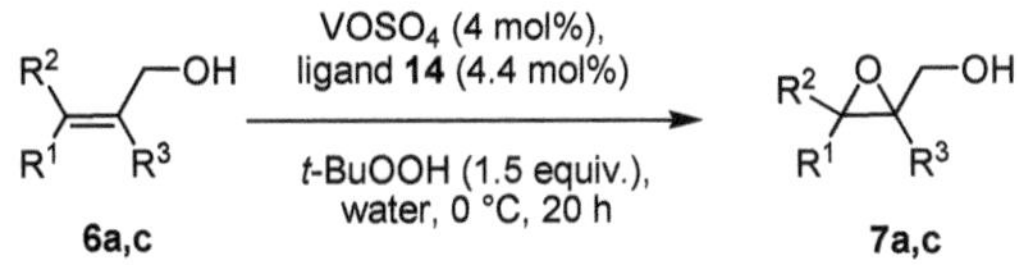

6a, 7a R^1 = Ph, R^2 = H, R^3 = CH_3
6c, 7c R^1 = n-C_3H_7, R^2 = R^3 = H

Scheme 6. Asymmetric epoxidation using ligand **14**.

As a result, we got only poor enantioselectivities (up to 49%), however, the ligand's reactivity seems to be quite reasonable, since we achieved 56% yield on substrate **6c** after only 20 h (entry 2, Table 2). This might be a good starting point for further improvements of the ligand, preserving the phosphinamide backbone and optimize both phosphine and amino acid moieties (see Future Aspects, Chapter 4).

entry	substrate	product	ee (%)	yield (%)
1	**6a**	(*S,S*)-**7a**	49 [a]	- [b]
2	**6c**	(*R,R*)-**7c**	47 [c]	56

Table 2. Results of epoxidations using ligand **14**. *a* detected by chiral GC (Supelco β-Dex 120 column; oven temp. 110 °C for 2 min, then 1 °C/min to 200 °C, $t_{S,S}$ = 34.66, $t_{R,R}$ = 35.27); *b* not determined; *c* Supelco β-Dex 120 column, oven temp. 70 °C for 2 min, then 1 °C/min to 200 °C, $t_{(S,S)}$ = 15.19, $t_{(R,R)}$ = 15.52.

3.3 Hydroxamic acids containing sulfinamide moiety

3.3.1 Synthesis of ligand 19 using direct method

A closer look at the results achieved with ligands **3** and **14** led us to the idea of improving the selectivity by introducing a stereogenic centre closer to the amino proton while maintaining the amide motif. We therefore chose chiral sulfinyl amides due to their similar reactivity and facile availability.

Starting from sodium *p*-toluenesulfinate, which was converted into corresponding acid chloride **15** using thionyl chloride[27], we tried to couple it directly with the free amino acid under aqueous conditions (Scheme 7). Following a literature protocol[6] related to sulfonyl compounds, we encountered hydrolysis of product **16** into sulfinic acid and amino acid when using concentrated hydrochloric acid (pH 1-2) for the reaction work-up. We then used citric acid (pH 4-5) due to this obvious acid lability of sulfinamide S-N bond, but still we could not isolate the expected racemic mixture of free acid **16**, maybe caused by poor solubility qualities of the crude product.

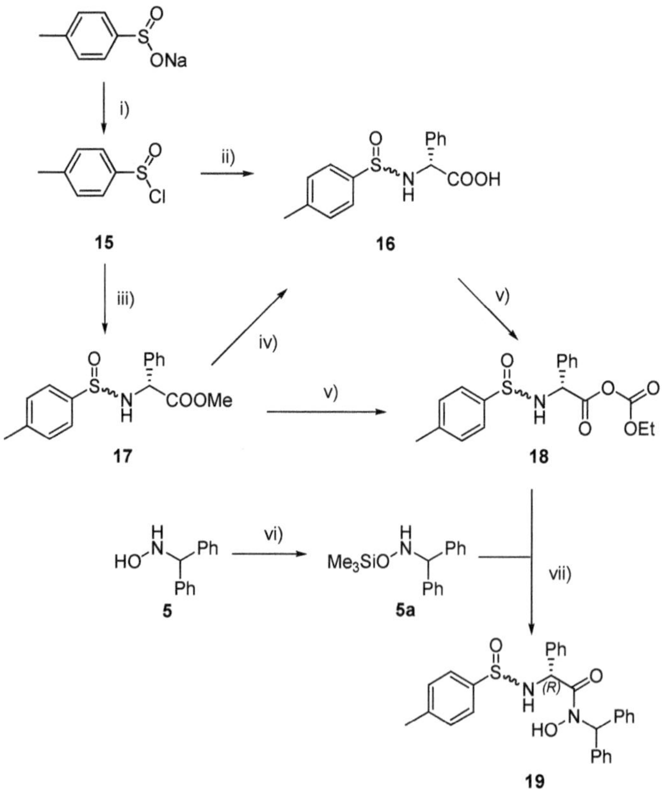

Scheme 7. Considered synthesis pathways to ligand **19** using direct method. Conditions: i) $SOCl_2$, r.t.; ii) (*R*)-phenylglycine, $NaOH_{aq.}$, Et_2O, r.t.; iii) (*R*)-phenylglycine methyl ester hydrochloride, DIEA, anh. DCM, r.t.; iv) $NaOH_{aq.}$, DCM, r.t.; v) ethyl chloroformate, NMM, anh. THF, -10 °C; vi) chlorotrimethylsilane, 2,6-lutidine, THF, 0 °C; vii) 2,6-lutidine, THF, r.t.

Continuing this synthesis pathway, it was planned to activate **16** using ethyl chloroformate to obtain the mixed anhydride **18**, which could then be converted into ligand **19** by coupling with *O*-protected hydroxylamine **5a**.

Exploring a different pathway, we were able to synthesize racemic methyl ester **17** starting from sulfinyl chloride **15** and (*R*)-phenylglycine methyl ester hydrochloride under anhydrous conditions. Unfortunately, subsequent mild alkaline hydrolysis of **17** to free acid **16** could not be accomplished successfully for some reason. As well, the *in situ*-activation of ester **17** into mixed anhydride **18** followed by coupling with beforehand trimethylsilyl-*O*-protected hydroxylamine **5a** did not afford the desired diastereomeric mixture of ligands (*R*, *R*)-**19** and (*S*, *R*)-**19**, but only hydroxylamine **5** was detected by TLC and ^{1}H-NMR.

3.3.2 Synthesis of ligand 19 using Nosyl protecting group

Considering the sensitivity of sulfinamides against acids, we decided to modify the *N*-protected amino acid first and introduce the sulfine moiety not before the last step. Starting from (*R*)-phenylglycine, we used a known procedure[32] to introduce the *o*-nitrophenylsulfonyl (Nosyl) moiety and could successfully isolate *N*-protected amino acid **20** (Scheme 8). Due to bad solubility qualities of this compound, further conversions turned out to be difficult. We tried to synthesize acid chloride **21** starting from acid **20** by using phosphorus pentachloride or thionyl chloride as chlorination reagent.[6] In the former case, we could only gain back the starting material, whereas in the latter case we were able to isolate crude acid chloride **21**. Subsequent coupling with *O*-protected hydroxylamine **5a**, however, did not lead to desired product **23** (Scheme 8).

A second effort was made to convert *N*-protected amino acid **20** into compound **23**, this time via *in situ*-activation as mixed anhydride **22** and followed by coupling with **5a**. In this case, we could not detect any product signals and we therefore decided not to pursue this synthesis pathway any more.

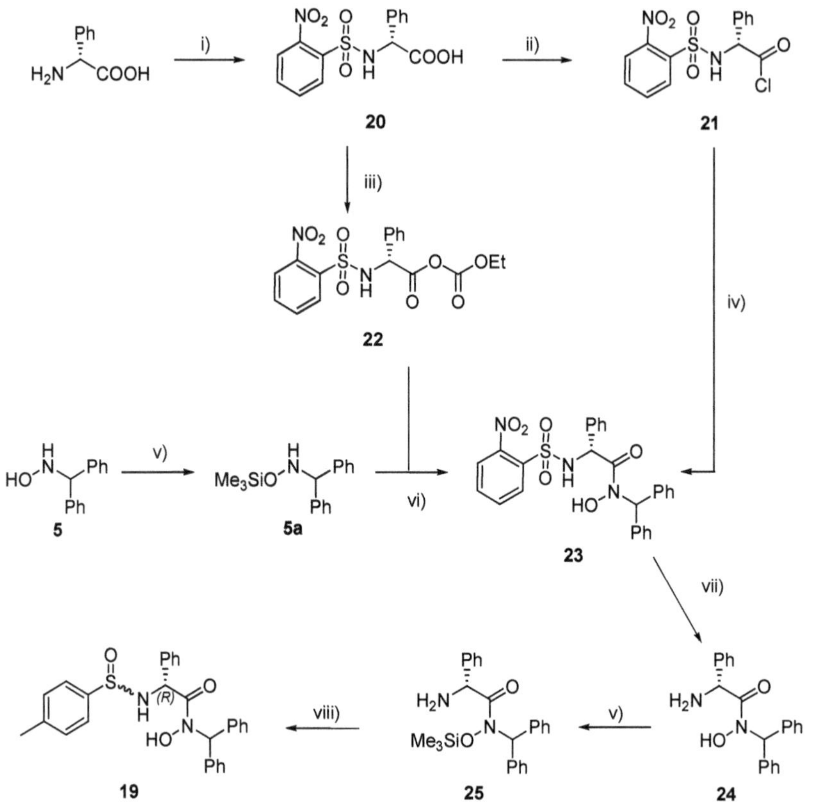

Scheme 8. Considered synthesis pathways to ligand **19** using the Nosyl protecting group. Conditions: i) *o*-Nitrophenylsulfonyl chloride, $NaOH_{aq.}$, Et_2O, 70-80 °C; ii) PCl_5, DCM, r.t. or $SOCl_2$, reflux; iii) ethyl chloroformate, NMM, anh. THF, -10 °C; iv) **5a**, 2,6-lutidine, THF, r.t.; v) chlorotrimethylsilane, 2,6-lutidine, THF, 0 °C; vi) 2,6-lutidine, THF, r.t.; vii) 2-Mercaptoethanol, LiOH, DMF, r.t.; viii) *p*-toluenesulfinyl chloride, 2,6-lutidine, THF, r.t.

To complete the synthesis, it would be necessary to remove Nosyl protecting group from compound **23**, which could be accomplished by literature procedure[32]. Subsequent *O*-protection and introduction of sulfine moiety would lead to desired ligand **19**, however, as a diastereomeric mixture which one would need to resolve.

3.4 Synthesis of *N*-Benzhydryl-hydroxylamine 5

All hydroxamic acids considered in this work contain the *N*-benzhydryl-hydroxylamine moiety as an essential part of the -C(O)N(OH)CHPh$_2$ structural motif (Figure 10). As hydroxylamines readily undergo N-C bond formation when using activated carboxylic acid derivatives as electrophiles, such as acid chlorides or mixed anhydrides, this particular coupling reaction represents the key step in synthesis pathways utilized to prepare ligands **3**, **14** and **19**.

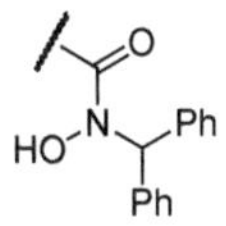

Figure 10. Structural motif of substituted hydroxamic acids.

Synthesis of hydroxylamine **5** was performed in three steps, starting from *N*-benzhydrylamine using a published protocol[29] (Scheme 9). The first step consisted of treating the amine with bromoacetonitrile in presence of Hünig's base (DIEA) to afford the cyanomethylated amine.[30] The following oxidation with *m*-CPBA furnished nitrone **4**, which was converted with hydroxylamine hydrochloride to give the desired *N*-benzhydryl-hydroxylamine (**5**).

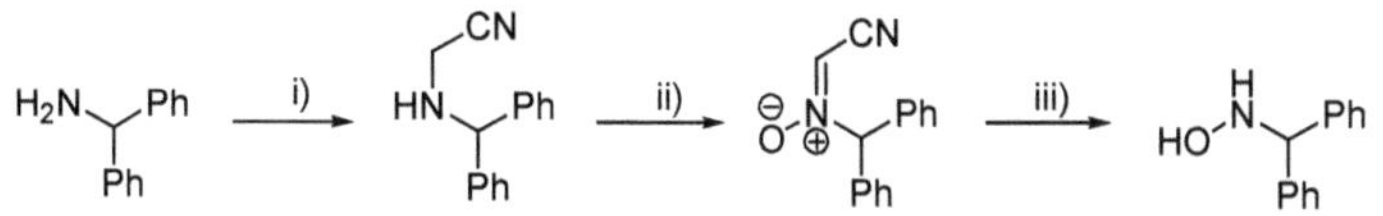

4 5

Scheme 9. Synthesis pathway of hydroxylamine **5**. Conditions: i) BrCH$_2$CN, DIEA, MeCN, reflux; ii) *m*-CPBA, DCM, 0 °C; iii) NH$_2$OH·HCl, MeOH, 60 °C.

4 Conclusions and Future Aspects

4.1 General

During this project, we reached the aim of establishing synthetic routes to new hydroxamic acid derivatives, which were used as ligands for asymmetric epoxidation of allylic alcohols in water. Three new ligands, (*S, R*)-**3**, (*S, S*)-**3** and **14** were successfully synthesized and subsequently tested on standard substrates. To accomplish the ligand synthesis, we used known procedures as well as new ones, for example the coupling of *in situ*-protected hydroxylamine **5a** and activated mixed anhydride **13**. As yields were not always satisfying (see Experimental), optimization of several steps could be a general aim for future work, especially considering the synthesis up-scaling of promising ligand **14**.

4.2 Camphor derivatives

Using Camphor-derived ligands **3**, we got enantiomeric excess values up to 46%, with comparible results for both diastereomers. This leads us to the conclusion that the Camphor moiety's intrinsic chirality does not sufficiently influence product formation regarding enantioselectivity. As well, the aliphatic scaffold is apparently not bulky enough to introduce additional selectivity as we expected before.

4.3 Sulfinamides

Several efforts to synthesize a ligand containing a chiral sulfur atom have not been successful. On the way to ligand **19**, we encountered various difficulties, especially caused by poor solubility properties. This is also why we finally abandoned the synthesis pathway using the Nosyl amino-protecting group. As it is reasonable to synthesize sulfinamide-based hydroxamic acids to apply them to epoxidation reactions, there remains the task of finding new, suitable synthetic approaches in a future project.

4.4 Phosphinamides

In order to amplify the steric extent, we successfully designed ligand **14** based on *N*-Dpp protected phenylglycine in a convergent five-step synthesis, carrying out similar reactions as for synthesizing hydroxamic acid **3.** It is the first time that a phosphinamide-derived hydroxamic acid has been used for asymmetric epoxidations in water. Although ligand **14** showed only enantiomeric excess values up to 49%, the phosphinamide moiety apparently causes reasonable reactivity concerning chemical yields.

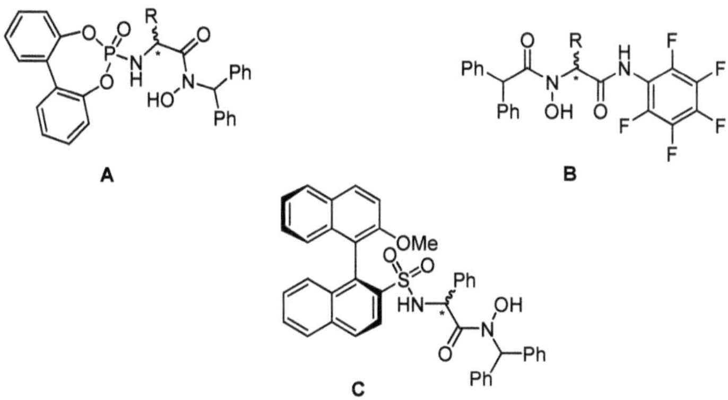

Figure 11. Possible future ligands containing different amide moieties.

In future work, it might be worth to try to modify the substituents on phosphorus and on the amino acid backbone in order to create a new stereogenic center on P as well as increasing the steric bulk (ligand **A**, Figure 11).

4.5 Anilides

Regarding the coordinating properties of the hydroxamic acids used as ligands so far, it is thought that interaction with vanadium takes place via carbonyl and hydroxyl oxygen atoms.[6] Furthermore, there is a possibility of the deprotonated amide nitrogen participating in coordination as a donor atom as well, limited by pK_a value of this proton. Hitherto used sulfonamides show pK_a values of about 9, and this is why it might be useful to try a pentafluoroanilide moiety instead of $NHSO_2R$, which would result in similar pK_a value (ligand **B**, Figure 11). A shortened possible retrosynthetic approach with pentafluoroaniline and (*S*)- or (*R*)-phenylglycine as starting materials is shown in Scheme 10.

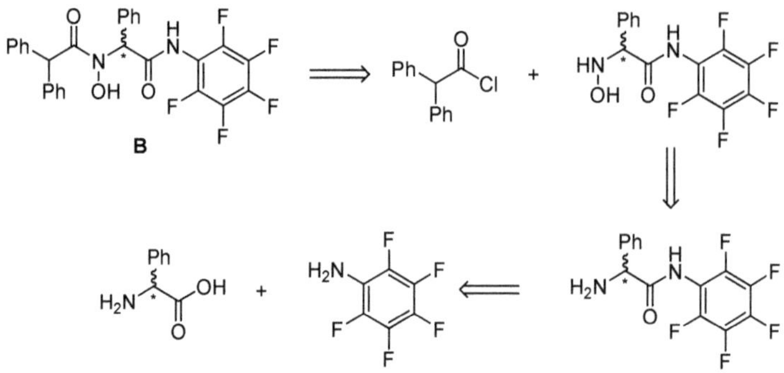

Scheme 10. Possible retrosynthetic approach to future ligand **B**.

4.6 Binaphthyls

Further investigations could then look at coupling a binaphthyl to the sulfonamide moiety. To combine amplification of sterical extent and to preserve chirality at the same time, binaphthylsulfonyl derivatives could be the compounds of choice. They both bear axial chirality and considerably bulky qualities, which might lead to improved selectivities in the epoxidation reactions.

A possible synthetic pathway, starting from enantiopure and easily available (*L*)-menthol is given in Scheme 12. By using (*L*)-menthyl esters in the Grignard-coupling step, one would avoid possible racemization of the formed axially chiral binaphthyl.[28] After a few more steps, including amination and chlorination of binapthyl moiety, one would introduce the amino acid backbone and the hydroxylamine moiety to gain the desired hydroxamic acid (ligand **C**, Scheme 12).

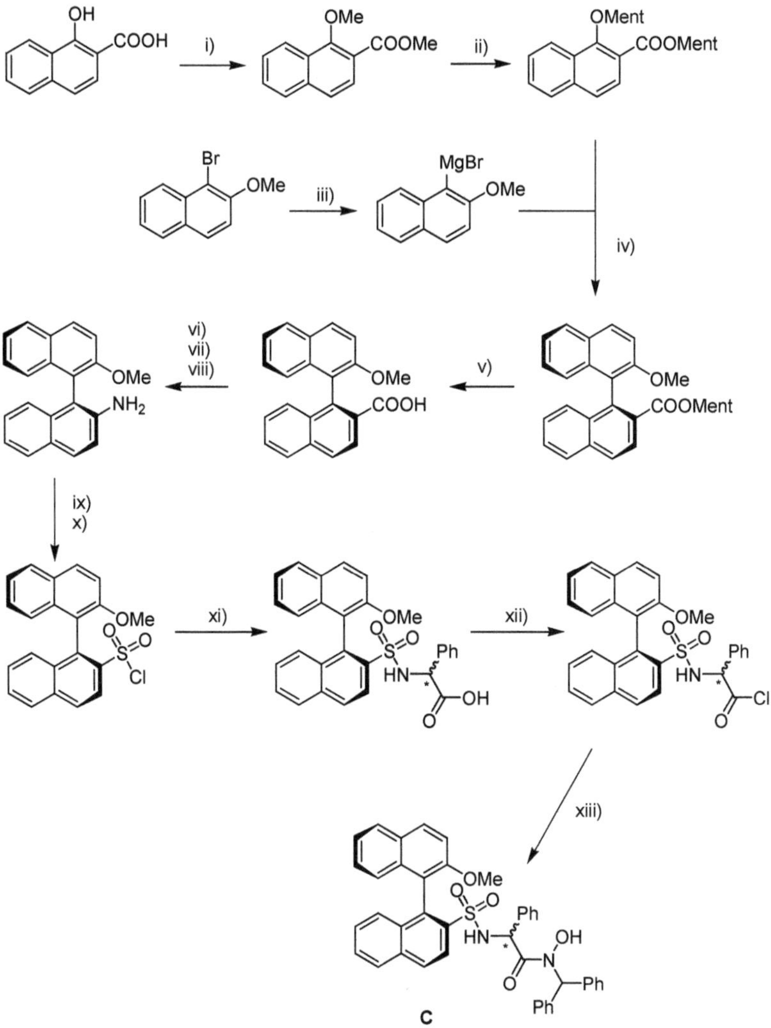

C

Scheme 11. Possible synthetic pathway to binaphthyl-derived ligand **C**. Conditions: i) NaH, MeI, DMF, 60 °C; ii) sodium (*L*)-menthoate, DMF, 50 °C; iii) Mg, Et_2O; iv) benzene: v) KOH / $EtOH_{aq}$; vi) Et_3N, ethyl chloroformate; vii) NaN_3; viii) ΔT, KOH; ix) $NaNO_2$, H_2SO_4; x) SO_2Cl_2: xi) (*S*)- or (*R*)-phenylglycine; xii) PCl_5; xiii) *N*-benzhydryl-hydroxylamine, TMSCl, 2,6-lutidine, 0 °C.

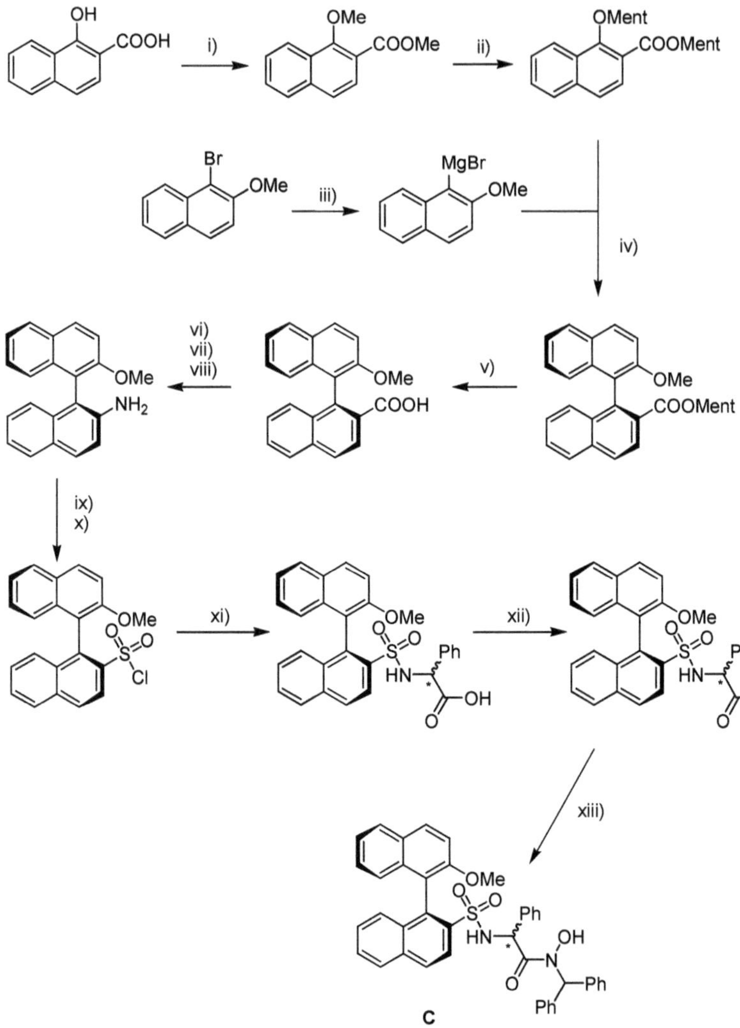

Scheme 12. Possible synthetic pathway to binaphthyl-derived hydroxamic acid ligand **B**. Conditions: i) NaH, MeI, DMF, 60 °C; ii) sodium (*L*)-menthoate, DMF, 50 °C; iii) Mg, Et_2O; iv) benzene: v) KOH / $EtOH_{aq}$; vi) Et_3N, ethyl chloroformate; vii) NaN_3; viii) ΔT, KOH; ix) $NaNO_2$, H_2SO_4; x) SO_2Cl_2: xi) (*S*)- or (*R*)-phenylglycine; xii) PCl_5; xiii) *N*-benzhydryl-hydroxylamine, TMSCl, 2,6-lutidine, 0 °C.

5 Experimental

5.1 General Methods

The NMR spectra were recorded in $CDCl_3$, ^1H at 400 MHz and ^{13}C at 100.6 MHz with chloroform-d_1 (δ 7.26, ^1H; δ 77.0, ^{13}C) as internal standard unless otherwise indicated.

The mass spectra (EI and/or CI) were measured on a dual sector mass spectrometer using direct inlet and the lowest temperature enabling evaporation.

Thin-layer chromatography reaction monitoring was accomplished using *Macherey-Nagel* ALUGRAM SIL G/UV_{254} TLC plates.

5.2 Synthesis of hydroxylamine

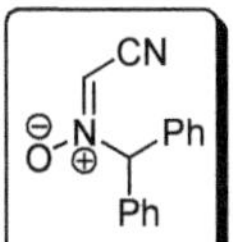

4

Nitrone 4. Following literature protocol[29], *m*-CPBA (8.96 g, 51.9 mmol) was dissolved in anh. DCM (70 mL), $MgSO_4$ (5 g) was added and the mixture stirred for 15 min. The solid was removed by filtration and washed with DCM (60 mL), then the solution was added slowly to a stirred solution of *N*-cyanomethyl-diphenylmethylamine[30] in anh. DCM (40 mL) at 0 °C (ice-bath). After addition was completed, cooling was removed and the mixture was stirred at r.t. for 2 h. Reaction mixture was then successively washed with aqueous saturated solutions of $NaHCO_3$ (2 x 100 mL), Na_2CO_3 (2 x 100 mL), and water (2 x 100 mL). Aqueous layer was extracted with DCM (60 mL and 40 ml), combined organic phase dried over $MgSO_4$ and solvent evaporated *in vacuo* to furnish 4.54 g (19.2 mmol, 92%) of **4** as a yellow solid, which was used without further purification. ^1H NMR (400 MHz, $CDCl_3$) δ 6.27 (s, 1H), 6.64 (s, 1H), 7.19-7.42 (m, 10H); NMR data consistent with the literature.[29]

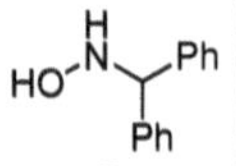

5

***N*-Benzhydryl-hydroxylamine (5).** To a stirred solution of nitrone **4** (4.45 g, 18.8 mmol) in methanol (80 mL) was added hydroxylamine hydrochloride (6.53 g, 94 mmol) and the mixture was stirred at 60 °C for 6.5 h. After cooling down to r.t., solution was concentrated *in vacuo*, DCM (50 mL) was added and the solution filtered through celite. The filtrate was washed with an aqueous saturated solution of Na_2CO_3 (2 x 50 mL) and the aqueous layer was extracted with DCM (50 mL). After drying over $MgSO_4$, solvent was evaporated *in vacuo* to afford 3.53 g (17.7 mmol, 94%) of **5** as a yellow-brown oil. ^{1}H NMR (400 MHz, $CDCl_3$) δ 5.33 (s, 1H), 7.28-7.41 (m, 10H); consistent with literature data.[29]

5.3 Synthesis of Camphor-derived hydroxamic acids

5.3.1 Sulfonamides

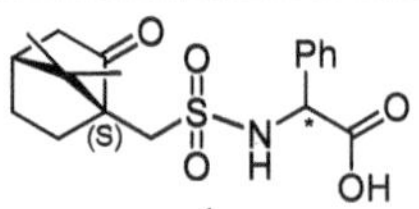

1

General procedure for the synthesis of sulfonamides 1. Following literature protocol[6], a solution of D-(*R*)-phenyl glycine or L-(*S*)-phenyl glycine (1 equiv.) and sodium hydroxide (2.5 equiv.) in water (3 mL) was stirred for 15 min at r.t. and a solution of (1*S*)-Camphor-10-sulfonyl chloride (1.2 eq) in diethyl ether (3 mL) was added and the mixture was stirred at r.t. overnight. Acidifying with 12 M HCl produced white precipitate, which was collected by filtration and washed with water.

***N*-((1*S*)-Camphor-10-sulfonyl)-(*R*)-phenyl glycine (*S*,*R*)-(1):** Yield: 190 mg, 0.52 mmol, 63%. ^{1}H-NMR (400 MHz, $CDCl_3$) δ 0.47 (s, 3H), 0.77 (s, 3H), 1.34 (m, 1H), 1.80–

1.99 (m, 5H), 2.25 (m, 1H), 2.59 (d, ^2J = 14.8 Hz, 1H), 2.72 (d, ^2J = 15.2 Hz, 1H), 5.24 (d, ^3J = 8.4 Hz, 1H), 6.64 (d, ^3J = 8.4 Hz, 1H), 7.29-7.38 (m, 5H).

***N*-((1*S*)-Camphor-10-sulfonyl)-(*S*)-phenyl glycine (*S,S*)-(1):** Was already made before current project. ^{1}H-NMR (400 MHz, $CDCl_3$) δ 0.92 (s, 3H), 1.01 (s, 3H), 1.47 (m, 1H), 1.93–2.04 (m, 5H), 2.44 (m, 1H), 2.99 (d, ^2J = 15.2 Hz, 1H), 3.59 (d, ^2J = 14.8 Hz, 1H), 5.37 (d, ^3J = 7.6 Hz, 1H), 6.66 (d, ^3J = 7.6 Hz, 1H), 7.38-7.47 (m, 5H).

5.3.2 Acid chlorides

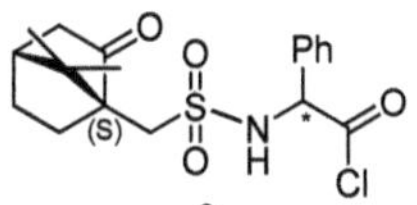

2

General procedure for the synthesis of *N*-protected acid chlorides 2. Following literature protocol[6], phosphorus pentachloride (1.2 equiv.) was added in one portion to a stirred suspension of *N*-protected amino acid **1** (1 equiv.) in anh. diethyl ether (5 mL) or solution in anh. DCM (5 mL) under argon. After stirring for 2 h at r.t., hexane (25 mL) was added and mixture was kept in freezer overnight. Precipitated white crystal needles were collected by filtration and washed with hexane.

***N*-((1*S*)-Camphor-10-sulfonyl)-(*R*)-phenyl glycine acid chloride (*S,R*)-(2):** Using diethyl ether. Yield: 57 mg, 0.16 mmol, 35%. ^{1}H-NMR (400 MHz, $CDCl_3$) δ 0.56 (s, 3H), 0.88 (s, 3H), 1.46 (m, 1H), 1.68–2.12 (m, 5H), 2.36 (m, 1H), 2.75 (d, ^2J = 15.2 Hz, 1H), 2.82 (d, ^2J = 15.2 Hz, 1H), 5.59 (d, ^3J = 8.8 Hz, 1H), 7.03 (d, ^3J = 8.8 Hz, 1H), 7.46 (s, 5H).

***N*-((1*S*)-Camphor-10-sulfonyl)-(*S*)-phenyl glycine acid chloride (*S,S*)-(2):** Using DCM. Yield: 259 mg, 0.72 mmol, >98%. ^{1}H-NMR (400 MHz, $CDCl_3$) δ 1.03 (s, 3H), 1.04 (s,

3H), 1.50 (m, 1H), 1.97–2.10 (m, 5H), 2.50 (m, 1H), 3.10 (d, ^{2}J = 14.8 Hz, 1H), 3.66 (d, ^{2}J = 15.2 Hz, 1H), 5.65 (d, ^{3}J = 8.4 Hz, 1H), 7.22 (d, ^{3}J = 8.4 Hz, 1H), 7.47 (m, 5H).

5.3.3 Hydroxamic acids

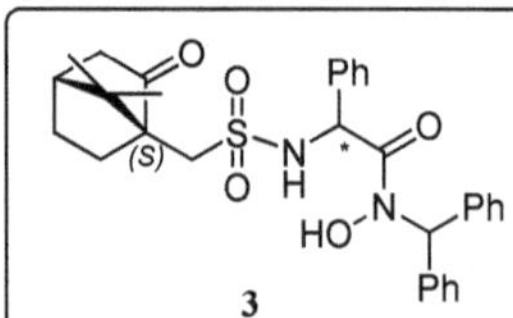

3

General procedure for synthesis of hydroxamic acids 3. Following literature protocol[31], to a stirred solution of hydroxylamine **5** (1 equiv.) in anh. THF (5 mL) was added 2,6-lutidine (2 equiv.) and trimethylsilyl chloride (2 equiv.) at 0 °C under argon, and reaction was monitored by TLC. After stirring for 4 h at r.t., mixture was cooled to 0 °C, acid chloride **2** was added in one portion and stirring continued at 0 °C for 15 min. Then mixture was warmed to r.t., stirred for 2 h and then water (0.5 mL) was added. Storring was continued at r.t. for 30 min , then THF was evaporated *in vacuo* and ethyl acetate (10 mL) was added to the residue. Organic layer was successively washed with each 15 mL 10% citric acid, 5% $NaHCO_3$ solution, water then dried (MgSO4), and evaporated *in vacuo* to afford a crude product which was purified by column chromatography (30 x 1.5 cm, ethyl acetate - PE 1:2) to give **3** as a white solid.

(*R*)-*N*-benzhydryl-2-(((1*S*,4*R*)-7,7-dimethyl-2-oxobicyclo[2.2.1]heptan-1-yl) methylsulfonamido)-*N*-hydroxy-2-phenylacetamide (*S*,*R*)-(3): Yield: 38 mg, 0.07 mmol, 73%. ^{1}H-NMR (400 MHz, $CDCl_3$) δ 0.48 (s, 3H), 0.81 (s, 3H), 1.33 (m, 1H), 1.76–1.98 (m, 5H), 2.13 (m, 1H), 2.55 (d, ^{2}J = 14.8 Hz, 1H), 2.86 (d, ^{2}J = 14.8 Hz, 1H), 5.75 (d, ^{3}J = 9.2 Hz, 1H), 6.46 (d, ^{3}J = 8.8 Hz, 1H), 6.83 (d, ^{3}J = 7.2 Hz, 1H), 7.09–7.36 (m, 15H), 7.98 (bs, 1H). ^{13}C-NMR (100 MHz, $CDCl_3$) δ 19.4 (CH_3), 19.7 (CH_3), 26.3, 27.0, 29.7, 42.6, 42.8, 48.5, 51.8, 58.0, 59.1, 63.1, 127.4-129.6, 136.1, 137.4, 137.7, 169.7, 216.1.

(S)-*N*-benzhydryl-2-(((1*S*,4*R*)-7,7-dimethyl-2-oxobicyclo[2.2.1]heptan-1-yl) methyl-sulfonamido)-*N*-hydroxy-2-phenylacetamide (*S*,*S*)-(3): Yield: 166 mg, 0.22 mmol, 44%. ^{1}H-NMR (400 MHz, $CDCl_3$) δ 0.71 (s, 3H), 0.86 (s, 3H), 1.18 (m, 1H), 1.57-2.00 (m, 5H), 2.26 (m, 1H), 2.76 (d, ^{2}J = 14.4 Hz, 1H), 3.34 (d, ^{2}J = 14.8 Hz, 1H), 5.81 (d, ^{3}J = 8.0 Hz, 1H), 5.96 (s, 1H), 6.89 (d, ^{3}J = 7.2 Hz, 1H), 7.12–7.36 (m, 15H). ^{13}C-NMR (100 MHz, $CDCl_3$) δ 19.4 (CH_3), 19.8 (CH_3), 26.2, 27.0, 42.8, 42.9, 48.6, 51.2, 57.9, 59.0, 63.2, 127.7-129.1, 136.5, 137.3, 137.5, 170.3, 216.4. HRMS (EI) *m/z* 546.2186 ($C_{31}H_{34}N_2O_5S$ requires 546.2188).

5.4 Synthesis of Dpp-derived hydroxamic acid

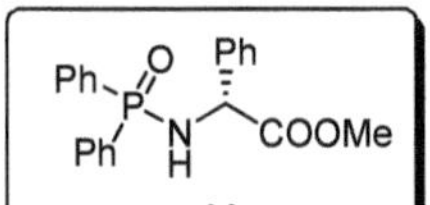

11

***N*-(diphenylphosphino)-(*R*)-phenylglycine methyl ester (11).** Following literature protocol[12], to a suspension of (*R*)-phenylglycine methyl ester hydrochloride (2.02 g, 10.0 mmol) in DCM (40 mL) was added NMM (2.02 g, 20.0 mmol) at 0 °C under argon. To the vigorously stirred mixture was then added a solution of diphenylphosphinic chloride (2.37 g, 10.0 mmol) in DCM (8 mL) dropwise via syringe. After stirring for 1 h in ice-bath, solvent was removed *in vacuo*. To the pale yellow residue was added ethyl acetate (15 mL) and water (15 mL), then organic layer was successively washed with saturated aqueous solution of $NaHCO_3$ (2 x 20 mL), 5% citric acid (20 mL + 10 mL), sat. $NaHCO_3$ (2 x 10 mL), water (2 x 10 mL) and brine (10 mL). After drying over $MgSO_4$, solvent was removed *in vacuo* to give crude Dpp-phenylglycine methyl ester (**11**). Yield: 2.56 g, 7.02 mmol, 70%. ^{1}H-NMR (400 MHz, $CDCl_3$) δ 3.61 (s, 3H), 4.24 (bs, 1H), 4.88 (t, *J* = 9.6 Hz, 1H), 7.17-7.47 (m, 10 H), 7.64-7.82 (m, 5H). ^{13}C-NMR (100 MHz, $CDCl_3$) δ 52.9, 56.8, 127.0, 128.2-128.8, 131.3-132.8, 138.3, 172.7. ^{31}P-NMR (162 MHz, $CDCl_3$) δ 24.1.

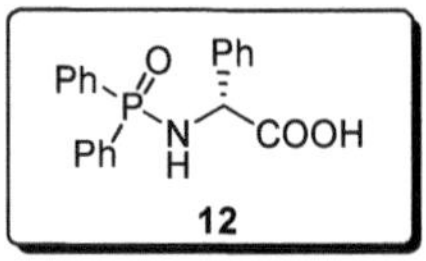

N-(diphenylphosphino)-(R)-phenylglycine (12). To a stirred solution of **11** (2.01 g, 5.50 mmol) in DCM (20 mL) was added sodium hydroxide (253 mg, 6.33 mmol) in water (3.3 mL) at r.t. After stirring for 35 h, DCM was removed *in vacuo* and aqueous layer was acidified with saturated citric acid solution to pH 3-4. The white precipitate was collected by filtration and washed with ethyl acetate to afford crude free acid **12**. Yield: 1.05 g, 2.99 mmol, 54%. ^{1}H-NMR (400 MHz, $CDCl_3$) δ 4.55 (dd, J = 3.2 Hz and 9.8 Hz, 1H), 4.82 (dd, J = 9.8 Hz and 12.8 Hz, 1H), 7.19-7.45 (m, 10 H), 7.60-7.83 (m, 5 H).

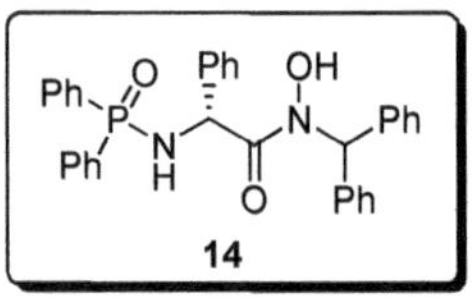

(R)-N-benzhydryl-2-(diphenylphosphorylamino)-N-hydroxy-2-phenylacetamide (14). To a stirred solution of hydroxylamine **5** (508 mg, 2.55 mmol) in anh. THF (5 mL) was added chlorotrimethylsilane (554 mg, 5.10 mmol) and 2,6-lutidine (547 mg, 5.10 mmol) at 0 °C under argon. The mixture was then stirred at r.t. for 4 h. Meanwhile, a suspension of acid **12** (984 mg, 2.80 mmol) in anh. THF (5 mL) was added NMM (283 mg, 2.80 mmol) and mixture stirred at –10 °C under argon for 30 min. Then ethyl chloroformate (304 mg, 2.80 mmol) was added and mixture was stirred at 0 °C for 1 h. The precipitated salt was filtered off quickly and the filtrate was immediately added to the first solution at 0 °C. Stirring was continued for 30 min at 0 °C, then at r.t. overnight. Water (0.5 mL) was added and after stirring for 30 min at r.t., THF was removed *in vacuo*. To the residue was added ethyl acetate (10 mL) and organic layer washed successively with 10% citric acid (10 mL), 5% $NaHCO_3$ aqueous solution (10 mL), water (10 mL) and dried over $MgSO_4$. Solvent was removed *in vacuo* to afford a red-brown oil, which was purified by column chromatography (30 x 1 cm, ethyl acetate - PE 1:2) to give ligand **14** in 15% yield (196 mg, 0.37 mmol). ^{1}H-NMR (400 MHz, $CDCl_3$) δ 4.88 (dd, J = 1.6 Hz and 3.6 Hz, 1H), 5.74 (dd, J = 8.6 Hz and 12.4 Hz, 1H), 6.76-7.45 (m, 26 H), 10.66 (bs, 1 H). ^{13}C-NMR (100 MHz, $CDCl_3$) δ 53.8, 63.0, 127.0-128.6, 129.9, 131.5-132.7, 138.1, 138.7, 171.1. ^{31}P-NMR (161 MHz, $CDCl_3$) δ 25.1.

5.5 Synthesis of sulfinamide-derived amino acid

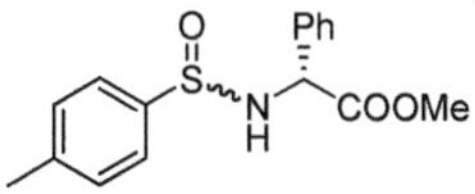

17

***rac-N-(p*-Toluenesulfinyl)-(*R*)-phenylglycine methyl ester (17).** To a stirred solution of D-(*R*)-phenylglycine methyl ester hydrochloride (333 mg, 1.65 mmol) and DIEA (640 mg, 4.95 mmol) in anh. DCM (10 mL) was added a solution of *p*-toluenesulfinyl chloride (342 mg, 1.96 mmol) under argon at r.t. After stirring overnight, water (20 mL) was added and the aqueous layer was extracted with DCM (3 x 10 mL). Combined organic phase was dried over $MgSO_4$ and solvent removed *in vacuo* to afford **15** as a yellow oil. Crude yield: 563 mg, > 98%. ^{1}H-NMR (400 MHz, $CDCl_3$) δ 2.27 (s, 3H), 2.35 (s, 3H), 3.56 (s, 3H), 3.64 (s, 3H), 4.77 (d, J = 8.0 Hz, 1H), 5.02 (d, J = 5.6 Hz, 1H), 5.07 (d, J = 5.6 Hz, 1H), 5.51 (d, J = 7.2 Hz, 1H), 6.84-7.55 (m, 18H).

5.6 Synthesis of *N*-Nosyl-(*R*)-phenylglycine

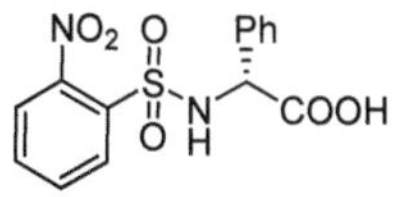

20

***N*-(*o*-Nitrophenylsulfonyl)-(*R*)-phenylglycine (20).** Following literature protocol[32], to a stirred solution of (*R*)-phenylglycine (3.02 g, 20.0 mmol) and sodium hydroxide (2.00 g, 50.0 mmol) in water (15 mL) was added a solution of *o*-nitrophenylsulfonyl chloride (4.43 g, 20.0 mmol) in diethyl ether (20 mL) and mixture was heated to 70 - 80 °C for 30 min. After cooling to r.t., mixture was acidified with 12 M hydrochloric acid to pH 1-2 precipitate a white solid, which was collected by filtration and washed with water (2 x 25 mL). Drying *in vacuo* afforded **15**. Yield: 2.397 g, 7.13 mmol, 36%. ^{1}H-NMR (400 MHz, DMSO / TFA) δ

5.04 (d, J = 9.2 Hz, 1H), 7.26-7.93 (m, 9H), 9.14 (d, J = 8.8 Hz, 1H). ^{13}C-NMR (400 MHz, DMSO / TFA) δ 55.5, 127.4-133.2, 169.8.

5.7 Epoxidation in water

General procedure for asymmetric epoxidation in water. Following literature protocol[5], vanadyl sulfate hydrate (3.3 mg, 20 µmol), ligand **3** or **14** (22 µmol) and allylic alcohol (1 mmol or 0.5 mmol) were added to distilled water (3 mL) at 0 °C. The mixture was stirred at room temperature for 30 min and then cooled to 0 °C. A 70% aqueous solution of *t*-BuOOH (1.5 mmol or 0.75 mmol) was added and the mixture was stirred at the same temperature for 48 h. The reaction mixture was then quenched with a concentrated solution of Na_2SO_3 (10 mL or 5 mL) and after stirring for 1 h at 0 °C it was extracted with DCM (2 × 20 mL), the combined organic extracts were dried over MgSO4 and concentrated *in vacuo* to give a brown oil. Purification of the products was accomplished by column chromatography on silica gel (30 × 1.5 cm) with ethyl acetate - PE mixture (1:2 or 1:4).

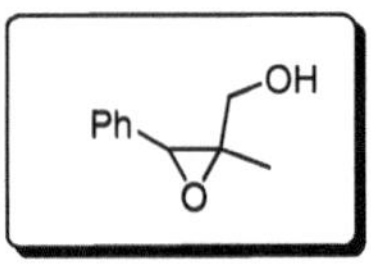

(2*S*, 3*S*)- 2-Methyl-3-phenyl-2,3-epoxy-propan-1-ol (7a): ^{1}H-NMR (400 MHz, CDCl3): δ = 1.12 (s, 3H), 2.16 (bs, 1H), 3.77 (d, J = 12.5 Hz, 1H), 3.88 (d, J = 12.5 Hz, 1H), 4.24 (s, 1H), 7.28-7.39 (m, 5H), consistent with the literature data.[33]

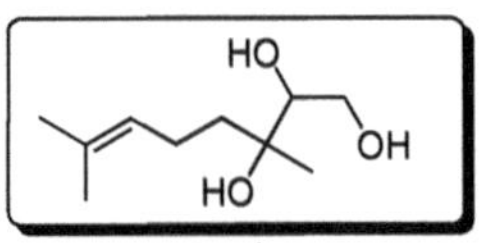

3,7-Dimethyloct-6-ene-1,2,3-triol (8): ^{1}H-NMR (400 MHz, $CDCl_3$) δ 1.24 (s, 3H), 1.37-1.45 and 1.56-1.64 (m, 2H), 1.62 (s, 3H), 1.69 (s, 3H), 2.08-2.15(m, 2H), 2.80 (bs, 1H), 3.45-3.50 (m, 1H), 3.73- 3.77 (m, 2H), 5.10-5.13 (m, 1H), consistent with the literature data.[34]

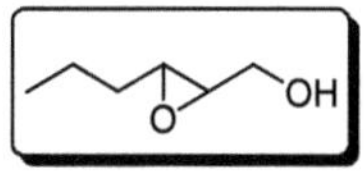

(*2R*, *3R*)-2,3-Epoxy-1-hexanol (7c) was isolated as a clear oil. Yield: 33 mg, 0.28 mmol, 56%. For GC analysis the product was converted into trifluoroacetate derivative by stirring with an excess (1.5 eq.) of trifluoroacetic anhydride in DCM at 20 °C for 1 h. ^{1}H-NMR (400 MHz, $CDCl_3$) δ 0.91 (t, J = 4.8 Hz, 3H), 1.34-1.52 (m, 4H), 2.88 (dt, J = 4.4, 2.8 Hz, 2H), 3.54 (dd, 12.8, 4.4 Hz, 1H), 3.85 (dd, 12.8, 2.8 Hz, 1H) consistent with literature data.[6]

6 References

1. Arends, I. W. C. E. *Angew. Chem. Int. Ed.* **2006**, *45*, 6250–6252.
2. Bolm, C. *Coord. Chem. Rev.* **2003**, *237*, 245–256.
3. Michaelson, R. C.: Palermo, R. E.; Sharpless, K. B. *J. Am. Chem. Soc.* **1977**, *99*, 1990–1992.
4. Hoshino, Y.; Yamamoto, H. *J. Am. Chem. Soc.* **2000**, *122*, 10452–10453.
5. Bourhani, Z.; Malkov, A.V. *Chem. Commun.* **2005**, 4592–4594.
6. Malkov, A.V.; Bourhani, Z.; Kocovsky, P. *Org. Biomol. Chem.* **2005**, *3*, 3194–3200.
7. Chelucci, G. *Chem. Soc. Rev.* **2006**, *35(12)*, 1230–1243.
8. Money, T. *Nat. Prod. Rep.* **1985**, *2(3)*, 253–289.
9. Oppolzer, W. *Tetrahedron* **1987**, *43(9)*, 1969–2004.
10. Kenner, G. W.; Moore, G. A.; Ramage, R. *Tetrahedron Lett.* **1976**, *40*, 3623–3626.
11. Sieber, P.; Iselin, B. *Helv. Chim. Acta* **1968**, *51(4)*, 622–628.
12. Ramage, R.; Hopton, D.; Parrott, M.J. *J. Chem. Soc. Perkin Trans. I* **1984**, 1357–1370.
13. Hulst, R.; Heres, H.; Peper, N. C. M. W.; Kellogg, R. M. *Tetrahedron Asymm.* **1996**, *7(5)*, 1373–1384.
14. Gamble, M. P.; Smith, A. R. C.; Wills, M. *J. Org. Chem.* **1998**, *63*, 6068–6071.
15. Li, K.-Y.; Zhou, Z.-H.; Wang, L.-X.; Zhao, G.-F.; Choi, M. C. K.; Zhou, Q.-L.; Tang, C.-C. *Heteroatom Chemistry* **2003**, *14(3)*, 288–291.
16. Pikul, S.; McDow Dunham, K.L.; Almstead, N. G.; De, B.; Natchus, M. G.; Anastasio, M. V.; McPhail, S. J.; Snider, C. E.; Taiwo, Y. O.; Chen, L.; Dunaway, C. M.; Gu, F.; Mieling, G. E. *J. Med. Chem.* **1999,** *42,* 87–94.
17. Mikolajczk, M.; Drabowicz, J.; Kielbasinski, P. *Chiral Sulphur Reagents*; CRC: Boca Raton, FL, USA, **1997**.
18. Pelissier, H. *Tetrahedron* **2007**, *63*, 1297–1330.
19. Fernandez, I.; Khiar, N. *Chem. Rev.* **2003**, *103*, 3651–3705.
20. Morton, D.; Stockman, R. A. *Tetrahedron* **2006**, *62*, 8869–8905.
21. Montanari, F. in *Organic Sulphur Chemistry* (Ed. Stirling, C. J. M.), Butterworth, London, **1975**.
22. Souers, A. J.; Owens, T. D.; Oliver, A. G.; Hollander, F. J.; Ellman, J. A. *Inorg. Chem.* **2001**, *40*, 5299–5301.

23. Davis, F. A.; Zhang, Y.; Andemichael, Y.; Fang, T.; Fanelli, D. I.; Zhang, H. *J. Org. Chem.* **1999**, *64*, 1403–1406.

24. Basch, H. in *The chemistry of sulphinic acids, esters and their derivatives* (Ed. Patai, S.), Wiley, Chichester, **1990**.

25. Nudelman, A. in *The chemistry of sulphinic acids, esters and their derivatives* (Ed. Patai, S.), Wiley, Chichester, **1990**.

26. Senanayake, C. H.; Krishnamurthy, D.; Lu, Z.-H.; Han, Z.; Gallou, I. *Aldrichim. Acta* **2005**, *38(3)*, 93–104.

27. Kurzer, F. *Org. Synth. Coll. Vol. 4*, 937. Preparation accomplished before this project.

28. Hattori, T.; Hotta, H.; Suzuki, T.; Miyano, S. *Bull. Chem. Soc. Jpn.* **1993**, *66*, 613–622.

29. Tokuyama, H.; Kuboyama, T.; Amano, A.; Yamashita, T.; Fukuyama, T. *Synthesis* **2000**, 1299–1304.

30. Thanks go to Louise Czemerys for preparation.

31. Jin, Y.; Kim, D. H. *Tetrahedron Asymm.* **1997**, *8(22)*, 3699–3702.

32. DeRuiter, J.; Borne, R. F.; Mayfield, C. A. *J. Med. Chem.* **1989,** *32,* 145–151.

33. Gao, Y.; Hanson, R.M.; Klunder, J. M.; Ko, S. Y.; Masamune, H.; Sharpless, K.B. *J. Am. Chem. Soc.* **1987**, *109*, 5765–5780.

34. Hanson, R. M.; *Tetrahedron Lett.* **1984**, *25*, 3783–3786.